FIRST GRADE GEOGRAPHY RIVERS AND LAKES OF THE WORLD

Speedy Publishing LLC
40 E. Main St. #1156
Newark, DE 19711
www.speedypublishing.com

Some rivers form
when lakes overflow.

A river is freshwater flowing across the surface of the land, usually to the sea.

Rivers flow in channels. The bottom of the channel is called the bed and the sides of the channel are called the banks.

Water from
a river can
come from rain,
melting snow,
lakes, ponds, or
even glaciers.

Lakes are large bodies of water that are surrounded by land and are not part of an ocean.

A lake usually contains freshwater but some can be saltwater.

Lakes form when water finds its way into a basin. Lakes must have a continual source of new water, otherwise they will eventually dry up.

Most lakes only last a few thousand years and then disappear.

CPSIA information can be obtained
at www.ICGtesting.com
Printed in the USA
BVHW062115060622
639006BV00012B/426